International Environmental Labelling

Vol.5 of 11
For All People who wish to take care of Climate Change
Maintenance & Cleaning Products: (All-purpose Cleaners,
Abrasive Cleaners, Powders. Liquids, Specialty Cleaners,
Kitchen, Bathroom, Glass and Metal Cleaners, Bleaches,
Disinfectants and Disinfectant Cleaners)

Jahangir Asadi

Vancouver, BC CANADA

> **Suggest an ecolabel**
> If you think that we missed a label and/or you are an ecolabelling body, please consider to submit for the next editions of our 11 Volumes International Eco-labelling Book series. Please send your details, and we'll review your suggestions. Our goal is to be as comprehensive as possible, so thank you for your help!
> info@TopTenAward.Net

Copyright © 2022 by Top Ten Award International Network.

All rights reserved. No part of this publication may be reproduced, distributed or transmitted in any form or by any means, including photocopying, recording, or other electronic or mechanical methods, without the prior written permission of the publisher, except in the case of brief quotations embodied in critical reviews and certain other noncommercial uses permitted by copyright law. For permission requests, write to the publisher, addressed "Attention: Permissions Coordinator," at the address below.

Published by: Top Ten Award International Network
Vancouver, BC **CANADA**
Email: Info@TopTenAward.net
www.TopTenAward.net

Ordering Information:
Quantity sales. Special discounts are available on quantity purchases by universities, schools, corporations, associations, and others. For details, contact the "Sales Department" at the above mentioned email address.

International Environmental Labelling Vol.5/J.Asadi—2nd ed.
ISBN 978-1-7775268-7-0

Contents

About TTAIN 10

Introduction 13

General principles of environmental labelling 20

Types of environmental labelling 24

Types I environmental labelling 28

Types II environmental labelling 48

Types III environmental labelling 54

All about 'Eco-friendly' Cleaning products 56

Which cleaning products are 'Eco-friendly' 57

Homemade cleaning products 60

Chemical - free Recipes for homemade cleaning products 61

Homemade cleaning products for Bathroom 63

Homemade cleaning products for Kitchen 65

Homemade cleaning products for Lundry room 68

How to disinfect 72

The future of cleaning 74

TTAIN Pioneers 78

Bibliography 85

Search by logos 92

Using Algae to Clean Wastewater 97

Environmental friendly photos 98

I dedicate this book to my wife, Mahnaz

Acknowledgements:

I wish to thank my committee members, who were more than generous with their expertise and precious time. I would like to acknowledge and thank the Top Ten Award International Network for allowing me to conduct my research and providing any assistance requested.

It should be noted that all the required permissions for using the logos and trade marks has been obtained to be published in this volume.

vinegar

soap

Natural cleaning detergent

lemon

About TTAIN

Top Ten Award International Network

Top Ten Award international Network (TTAIN) was established in 2012 to recognize outstanding individuals, groups, companies, organizations representing the best in the public works profession.

TTAIN publishing books related to international Eco-labeling plans to increase public knowledge in purchasing based on the environmental impacts of products.

Top Ten Award International Network provides A to Z book publishing services and distribution to over 39,000 booksellers worldwide, including Apple, Amazon, Barnes & Noble, Indigo, Google Play Books, and many more.

Our services including: editing, design, distribution, marketing
TTAIN Book publishing are in the following categories:

Student
Standard
Business
Professional
Honorary

We focus on quality, environmental & food safety management systems , as well as environmnetal sustain for future kids. TTAIN also provide complete consulting services for QMS, EMS, FSMS, HACCP and Ecolabeling based on international standards.

ISO 14024 establishes the principles and procedures for developing Type I environmental labelling programmes, including the selection of product categories, product environmental criteria and product function characteristics, and for assessing and demonstrating compliance. ISO 14024 also establishes the certification procedures for awarding the label.

TTAIN has enough experiences to help create new ecolabeling programmes in different countries all over the world.
For more detail visit our website : http://toptenaward.net
and/or send your enquiery to the following email:
info@toptenaward.net

CHAPTER 1

Introduction

This book is dedicated to the subject of environmental labels. The basis for the classification of its parts goes back to the types of environmental labelling according to the classifications provided by the International Organization for Standardization. In each section, while presenting the relevant definitions, I mention the existing international standards and present examples related to each type of labelling. Environmental labelling is an important and significant topic, and its richness is added to every day, which has attracted the attention of many experts and researchers around the world. The idea of compiling this book, came to my mind when I observed that national environmental labelling models have been developed in most countries of the world, but in many other countries, the initial steps have not been taken yet. Therefore, I decided to create the first spark for the development of environmental labelling patterns in other countries by collecting appropriate materials and inserting samples of labelling patterns of different countries of the world. It should be noted that the description of each environmental label in this book does not indicate their approval or denial; they are included only to increase the awareness of all enthusiasts and consumers of the meanings and concepts derived from such labels. We hereby ask all interested parties around the world who wish to start an environmental labelling program in their country to benefit from our intellectual assistance and support in the form of consulting contracts. Increasing human awareness of the urgent need to protect the environment has led to changes in all levels of activities, including the production of marketing products, consumption, use, and sale of goods and services at the national and international levels. Stakeholders involved in environmental protection include consumers, producers, traders, scientific and technological institutes, national authorities, local and international organizations, environmental gatherings, and human society in general. Decisions by consumers and sellers of products are made not only on the basis of key points such as quality, price, and availability of

products but also on the environmental consequences of products, including the consequences that a product can have before, after and during production. The most important environmental consequences include water, soil, and air pollution along with waste generation, especially hazardous waste. Further consequences include noise, odor, dust, vibration, and heat dissipation as well as energy consumption using water, land, fuel, wood, and other natural resources. There are further effects on certain parts of the ecosystem and the environment. In addition, the environmental consequences not only include the natural use of the products but also abnormal and even emergency or accidental uses. The basis of studies and studies in this field is done through product life cycle evaluation, which generally involves the study and evaluation of environmental aspects and consequences of a category (product, service, etc.) because of the preparation of raw materials for production until they are used or discarded. Sometimes the phrase "review from cradle to grave" is used for such an evaluation. In addition to the above, the environmental consequences that may occur at any stage of the product life cycle, including the preliminary stages and its preparation, production, distribution, operation, and sale, should also be considered when evaluating it. This type of evaluation refers to product life cycle analysis from an environmental point of view,"

which is a useful tool for measuring the degree of environmental health of a product, comparing different products, improving product quality, and confirming the environmental health claims of the product. The environmental health analysis tool for products and services facilitates their placement in domestic or foreign markets, considering that the awareness of consumers and retailers about the environmental consequences of the product has increased, as has the accurate and explicit measurement by the people in charge at all levels. Local, national, and international in the field of environmental protection. Products that can claim to be environmentally complete in all stages of their life cycle and meet the mandatory and optional environmental needs are considered successful products. Environmental messages refer to the policies, goals, and skills of product manufacturing companies as part of the environmental management systems in which they are applied, and consumers and retailers are increasingly paying attention to this issue when making purchasing decisions. In addition, companies have been encouraged and even forced to adapt their environmental management systems to agencies and retailers and to local, national, international, and other environmental issues.

The environmental health message of a product can be conveyed to the consumer in various ways, including implicitly or explicitly. For example, the implicit or implicit message conveyed directly by the product to the customer is that the product is suitable for the intended use and purpose, and, without material waste in size, weight, and dimensions, is perfectly proportioned and without additional packaging. Sometimes it is necessary to convey these messages and claims about the correctness of the product quite clearly through magazines or other media as well as through certificates that are accurate, simple, and convincing to the consumer in the form of a label. These messages must be accurate and fact-based; otherwise they will nullify the product and create contradictory effects. Confirmation of these claims by a third-party organization will increase its credibility. It should also be noted that the multiplicity of these messages, depending on the type of products or companies producing them, confuses consumers in the market and also creates artificial boundaries or causes a differentiated distinction against certain products or companies. Various models, principles, and methods have been provided by local, regional, national, and international organizations to demonstrate product life cycle analysis and other guidelines on environmental management systems and their labels. At the national level, significant advances have been made in the design of environmental labels in various countries, including developing countries and the Scandinavian countries. For example, the first project was designated in Germany as a Blue Angel in 1977, later on Canada in 1988, the Scandinavian countries and Japan in 1989, the United States and New Zealand in 1990, India, Austria, and Australia in 1991, And in 1992, Singapore, the Republic of Korea, and the Netherlands developed their national environmental labelling. Environmental labels are an environmental management tool that is the subject of a series of ISO 14000 standards. These environmental labels provide information about a product or commodity in terms of its broad environmental characteristics, whether it is about a specific environmental issue or about other characteristics and topics.Interested and pro-environmental buyers can use this information when choosing products or goods. Product makers with these environmental labels hope to influence people's purchasing decisions. If these environmental labels have this effect, the share of the product in question can increase, and other suppliers may create healthy environmental competition by improving the environmental aspects of their products and commodities. The overall goal of environmental labels is to convey acceptable and accurate information that is in no way misleading regarding the environmental aspects of products and commodities, and they encourage the consumer to buy and produce products that reduce stress on the environment. Environmental labelling must follow the general principles that the International Organization for Standardization has published in a collection entitled the ISO 14020 standard, which refers to these general principles here. It should be noted that other documents and laws in this field are considered if they are in accordance with the principles set out in ISO 14020.

Ecolabelling is a voluntary method of environmental performance certification and labelling that is practised around the world. An ecolabel identifies products or services proven to be environmentally preferable within a specific category.

CHAPTER 2

General Principles on Environmental Labelling

1 The First Principle: Evironmental notices and labels must be accurate, verifiable, relevant, and in no way misleading and/or deceptive.

2 The Second Principle: Procedures and requirements for environmental labels will not be ready for selection unless they are implemented by affecting or eliminating unnecessary barriers to international trade.

3 The Third Principle: Environmental notices and labels will be based on scientific analysis that is sufficiently broad and comprehensive, and to support this claim, the product must be reliable and reproducible.

4 The Fourth Principle: The process, methodology, and any criteria required to support the announcements on environmental labels will be available upon request all interested groups.

The Fifth Principle: Development and improvement of environmental notices and labels should be considered in all aspects related to the service life of the product.

The Sixth Principle: Announcements on environmental labels will not prevent initiative and innovation but will be important in maintaining environmental implementation.

The Seventh Principle: Any enforcement request or information requirement related to environmental notices and labels should be limited to the necessary information to establish compliance with an acceptable standard and based on the notification standards and environmental labels.

The Eighth Principle: The process of improving the announcement and environmental labels should be done by an open solution with interested groups. Reasonable impressions must be made to reach a consensus through this process.

The Ninth Principle: Information on the environmental aspects of the product and goods related to an advertisement and environmental label will be prepared for buyers and interested buyers from a group consisting of an advertisement and an environmental label.

What makes cleaning products eco-friendly?
Not containing toxic chemicals.
Has biodegradable ingredients.
Being made from plants.
Not using single use plastic.
Being plastic free entirely.
Being refillable.
Not creating waste.
Being vegan and/or cruelty free.

ECO HO
nature fri

SEHOLD
y products

CHAPTER 3

Types of Environmental Labelling

At present, according to the classification provided by the International Organization for Standardization, there are three types of environmental labelling patterns:

1. Type I labelling: This labelling is known as eco-labelling, and because it is difficult to translate this word into many languages, it presents another reason to adhere to a numerical classification system. In the content of Type I labelling, a set of social commitments that creates criteria according to the scientific principles on the basis of which a product is environmentally preferable is discussed. Consumers are then instructed in assessing environmental claims and must decide which packaging is more important.

2. Type II labelling: refers to the claims made on product labels in connection with business centers. This includes familiar claims such as recyclable, ozone-friendly, 60% phosphate-free, and the like. This type of labelling can be in the form of a mark or sentence on the product packaging. Some of them are valid environmental claims—and some can be completely misleading. Usually, all countries have laws against deceptive advertisements, so why has the International Organization for Standardization discussed this issue? The answer is that it is not clear whether the environmental claims have a technical basis or whether the ad is meaningless.

3 Type III labelling: is a distinct form of third-party environmental labelling pattern designed to avoid the difficulties that can result from type-one labelling. Technical committee for Environment of International organization for Standardization has undertaken a new project to standardize guidelines and Type III labelling methods. One of the main objections raised by industries to Type I labelling is the basis for its management.

Eco-Friendly Products Are Cost-Effective
In contrast, green cleaning products are less abrasive and offer cost saving opportunities such as reduced cost on repair and replacement of damaged floors and surfaces, safer work environments, and reduced water and chemical use

CHAPTER 4

Type I Environmental Labelling

Type I labelling: This labelling is known as eco-labelling, and because it is difficult to translate this word into many languages, it presents another reason to adhere to a numerical classification system. In the content of Type I labelling, a set of social commitments that creates criteria according to the scientific principles on the basis of which a product is environmentally preferable is discussed. Consumers are then instructed in assessing environmental claims and must decide which packaging is more important.

Type I adhesive has the following specifications:
A. Has an optional third-party template.
B. When the product meets a certain standard, the labelling of this product is included.
C. The purpose of this program is to identify and promote products that play a pioneering role in terms of environment, which means its criteria are at a higher level than the average environmental performance.
D. Acceptance/rejection criteria are determined for each group of products and are publicly available.
E. The criteria are adjusted after considering the environmental consequences of the product life cycle.

Examples of Type I Labelling:
In this section, and considering the importance of this type of labelling, I provide a description of some examples of Type I labelling related to some countries along with a list of products on which this mark is placed.

Global

The Ecological cleaning products certification enables your natural or organic cleaning products to be commercialized worldwide. Ecological cleaning products certification guarantees:
- environmentally friendly production and processing processes
- promotion of the use of natural or organic ingredients
- responsible management of natural resources
- prohibition of most of petrochemical ingredients

It is the certification through an independent certification body like Ecocert which enables to display the mention 'Natural detergent' with the label. 2 levels of labelling:
For the label Natural Detergents:
- highlighting of all the ingredients from natural origin:
- Maximum 5% of synthetic ingredients among restrictive list
- No environmental risky statement is authorized on the product

For the label Natural detergents made with Organic:
- Minimum 95% of ingredients are from natural origin
- Minimum 10% of ingredients are organic
- No risky statement is authorized on the product

Contact:
Web: https://www.ecocert.com/en-CA/
Tel.: (+1) 418-838-694

Russia

The Ecological Union is one of the leading Russian noncommercial organizations in the field of environmental protection. We have been conducting extensive work on environmental education and improving the environmental culture of society since 1991. Since 2001 we have been developing green standards, enhancing the production and consumption of environmentally preferable products. The main project of Ecological Union is the development of life-cycle based ecolabel (ISO 14024) Vitality Leaf, aimed to preserve a healthy environment for future generations.

This label on the packaging is a guarantee of the product environmental performance for people and the planet. It is awarded to companies who a leaders in sustainable production in its product group. As concerns about climate change, toxicity and waste are growing, those making responsible purchasing decisions need products with genuine environmental benefit. Life cycle-based ecolabels offer a viable market solution.

The ecolabel is based on a multi attribute criteria that are developed considering the entire product life cycle (Life Cycle Assessment approach). Criteria set thresholds values, limits and performances that are evaluated and certified by the Vitality Leaf experts. The Ecological Union acts in accordance with international practice and the official UN policy in the field of sustainable production and consumption. Our team consists of experts and auditors with professional environmental education, more than 10 years' experience in environmental certification, and leading auditors of international qualification (ISO 9001, ISO 14001, OHSAS 18001).

The Ecological Union has developed 28 standards. As of July 2021, 25 companies have received ecolabel for 180 products. The majority of them are in the following product groups: household chemicals, building materials, paints and varnishes, food and textile products.

Contact:
Phone / Fax: +7 (812) 571-38-38
191002, St. Petersburg, Rubinstein Street, 15-17, Premise 70 N, office 132
Email: mail@ecounion.ru

New Zealand

Environmental Choice New Zealand (ECNZ) is the country›s only Government-owned ecolabel. Administered by the New Zealand Ecolabelling Trust, the ecolabel was established in 1992 to provide a credible and independent guide for businesses and consumers to purchase and use products that are better for the environment.

A member of the Global Ecolabelling Network, ECNZ is a Type I ecolabel, which means products and services bearing the label meet criteria covering the whole life-cycle of the product/service, from raw materials, through manufacture and usage, to end-of-life disposal or reuse. Licensed products and services are independently assessed regularly by a third party.

The New Zealand Ecolabelling Trust
PO Box 56 533, Dominion Rd, Mt Eden, Auckland 1446
Tel: 0064 9 845 3330
Email: info@environmentalchoice.org.nz
Web: www.environmentalchoice.org.nz

China

China Environmental United Certification Center (CEC), approved by the Ministry of Ecology and Environment of the People's Republic of China (MEE) and accredited by Certification and Accreditation Administration Committee of PRC, is a comprehensive certification and service institution leading in environmental protection, energy saving and low carbon areas. . CEC is committed to serve building national ecological civilization; and has carried out research on environmental protection, energy saving, low carbon development strategies and solutions; has been continuously improving and innovating green industry evaluation system on industrial green development and transition CEC is building a bridge between green production and green consumption by offering independent, impartial and high-quality evaluation and certification service for government, enterprises and the public. CEC is a state-owned, non-profit, legal entity of independent third-party certification. It integrates the certification resource from the former National Accreditation Center for Environmental Conformity Assessment, the Secretariat of China Environmental Labelling Products Certification Committee, Environmental Development Center of MEE, the Chinese Research Academy of Environmental Sciences and other institutions. Business areas includes: products certification, management systems certification, services certification, addressing climate change, energy-saving and energy efficiency certification, green supply chain assessment, environmental stewardship, green credit assessment and green manufacturing system evaluation. CEC also carries out standard establishment and research project and international cooperation and exchanges, etc.

Contact:
Website: http://en.mepcec.com/
E-mail: zhangxiaoh@mepcec.com , zhangxiaoh@mepcec.com

Bolivia

Legally established in Bolivia, IMOcert has a presence in more than 20 countries in Latin America and the Caribbean, has regional offices in Peru, Paraguay, Mexico, among others. As an organic control body, it has been accredited for many years in accordance with the NOP / USDA Regulation, it also has accreditation of the ISO / IEC 17065 standard and also has other national accreditations of countries where it operates and authorizations for other sustainable schemes and social. IMOcert has extensive experience in certification of producer groups, actively collaborating in the origin and development of the internal control systems methodology.

Complete contact detail
Nombre/Name: Alberto Levy
Gerente Ejecutivo / Executive manager
Teléfono de oficina/office pone: (+591) 4456880/81
Fax: (+591) 44456882
Correo electrónico/ e-mail: imocert@imocert.bio – alevy@imocert.bio
www.imocert.bio

Hong Kong

The Green Council is a non-profit, tax-exempt charitable environmental stewardship organisation and certification body (Reg. No.: HKCAS-027) of Hong Kong established in 2000. A group of individuals from different sectors of industry and academics shared the vision to help build Hong Kong into a world-class green city for the future. They formed the Green Council with the aim of encouraging the commercial and industrial sectors to include environmental protection in their management and production processes. The Green Council is a non-profit, tax-exempt charitable environmental stewardship organisation and certification body (Reg. No.: HKCAS-027) of Hong Kong established in 2000. A group of individuals from different sectors of industry and academics shared the vision to help build Hong Kong into a world-class green city for the future. They formed the Green Council with the aim of encouraging the commercial and industrial sectors to include environmental protection in their management and production processes. The Green Council is a non-profit, tax-exempt charitable environmental stewardship organisation and certification body (Reg. No.: HKCAS-027) of Hong Kong established in 2000. A group of individuals from different sectors of industry and academics shared the vision to help build Hong Kong into a world-class green city for the future. They formed the Green Council with the aim of encouraging the commercial and industrial sectors to include environmental protection in their management and production processes.

Contact:
Website: https://www.greencouncil.org/hkgls
Email: info@greencouncil.org
Telephone: (852) 2810 1122

Peru

BIO LATINA, the consolidated byproduct of four Latin American national certification entities. Since 1998, we have provided certification services in Latin America for national and international markets. We seek to help create a more sustainable and resilient world. With these goals in mind, we have expanded our service portfolio beyond organic to social and environmental certifications.

Visit us: https://biolatina.com

From our regional offices we serve Latin American.

Our headquaters:
Av. Javier Prado Oeste 2501, Bloom Tower Of. 802, Magdalena del Mar, Lima 17, Perú

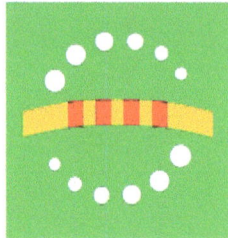

Catalonia

The Emblem of Guarantee of Environmental Quality identifies the products and services that have passed strict environmental quality criteria that go beyond regulatory requirements and bear in mind the life cycle. This type I ecolabelling system, adapted to ISO 14024, is compatible and on a par with other international ecolabelling systems such as the EU Ecolabel.

The Emblem of Guarantee of Environmental Quality was created in November 1994. Its original scope was guaranteeing the environmental quality of certain product properties and characteristics. In 1998, the scope was expanded to include services.

Through the creation of this ecolabel, Catalonia is eager to lead the way in terms of having its own regional ecolabelling system in Europe in keeping with European countries with a long history in environmental protection.

The purpose of the ecolabel is to encourage the design, production, marketing and consumption of more environmentally friendly products and services.

Contact details
Contact person: Josep M. Masip
josepmaria.masip@gencat.cat
ssq.tes@gencat.cat

Republic of Korea

The Korea Eco-labelling is a certification system enforced by the Ministry of Environment and KEITI(Korea Environmental Industry & Technology Institute). Since its foundation in April 1992, the system has certified a wide range of eco-friendly products, which were selected as excellent not only in terms of their environmental-friendliness, but also for their quality and performance during their life cycle. Korea Eco-labelling is voluntary certification scheme to attach logo to products with superior environmental quality throughout their lifecycle to other products of the same use, and thus to provide product information to consumers. For 30 years, the scheme has launched plenty of eco-labelling product standards covering personal and household goods, construction materials, office equipment furniture, etc. It products categories which cover all aspects of products, such as reduction of use of harmful substances, energy saving, resource saving, etc. As of April 30th 2021, 169 criterias(=standards), and certifications for 18,250 products(4,549 companies) have maintained.

Contact:
Korea Environmental Industry & Technology Institute(KEITI)
Office of Korea Eco-Label Innovation
Address: 215, Jinheung-ro, Eunpyeong-gu, Seoul, Repulic of Korea
T: +82 2 2284 1518
F: +82 2 2284 1526
E: accolly@keiti.re.kr
W: www.keiti.re.kr

USA

The Carbonfree® Product Certification is a meaningful, transparent way for you to provide environmentally-responsible, carbon neutral products to your customers. By determining a product's carbon footprint, reducing it where possible and offsetting remaining emissions through our third-party validated carbon reduction projects, companies can:
- Differentiate their brand and product
- Increase sales and market share
- Improve customer loyalty
- Strengthen corporate social responsibility & environmental goals

The Carbonfree® Product Certification Program is proud to be part of Amazon's Climate Pledge Friendly Program!
Carbonfund.org is leading the fight against climate change, making it easy and affordable to reduce & offset climate impact and hasten the transition to a clean energy future.

Contact:

O: 240.247.0630 ext 633
C: 203.257.7808
M: 853 Main Street, East Aurora, NY, 14052

Netherland

For more than 25 years, the independent Dutch foundation SMK works from professional knowledge with companies to improve the sustainability of products and business management. SMK cooperates with an extensive stakeholder network of governments, producers, branch and non-governmental organisations, retailers, consultancies, researchers. The SMK Boards of Experts establish objective criteria for more sustainable products and services. SMK's transparent work processes, third party audits and certifications are conducted according to international certification standards, mostly under supervision of the Dutch Accreditation Council. Besides, SMK is Competent Body of the EU Ecolabel. SMK keeps an extensive database of sustainability criteria.

Contact:
Bezuidenhoutseweg 105 - 2594 AC Den Haag
Telefoon: 070-3586300
Mobiel: 06-82311031
(niet op woensdag)
www.smk.nl

Taiwan

The Green Mark GM) Program was launched by the Environmental Protection Administration of Taiwan (TEPA) in 1992. As the official Type I eco-labeling program, it is in compliance with the requirements of the international stadard, ISO 14024 and is considered an important tool to promote green consumption and production .

To improve the GM application/review mechanism and introduce a third party certification scheme, TEPA promulgated the «Guideline for the Management of Certification Organizations for Environmental Protection Products" in June 2012. Both Environment and Development Foundation (EDF) and the Taiwan Testing and Certification Center (ETC) were commissioned by TEPA as official certifiers. With the expansion of certification capacity and authorization of the certification decision, the certification time was greatly reduced.

Contact :

Website: www.edf.org.tw
TEL: 886-3-5910008 #39
E-mail: lhliu@edf.org.tw

Denmark, Finland, Norway, Iceland, Sweden

The Nordic Swan Ecolabel
The Nordic Swan Ecolabel is the official Nordic ecolabel supported by all Nordic Governments. It is among the world›s strictest and most recognised environmental certifications.
The Nordic Swan Ecolabel is a Type I environmental labelling program established in 1989 by the Nordic Council of Ministers, connect¬ing policy, people, and businesses with the mission to make it easy to make the environmentally best choice. Nordic Ecolabelling is the non-profit organisation responsible for the Nordic Swan Ecolabel.
The organisation offers independent third-party certification and support for a wide range of product areas and services, ensuring that they comply with the Nordic Swan Ecolabel's strict requirements through documentation and inspections.

30 years of experience and expertise has made the Nordic Swan Ecolabel a powerful tool that paves the way to a sustainable future by giving producers a recipe on how to develop more environmentally sustainable products, and giving consumers credible guidance by helping them identify products that are among the environmentally best.

Globally, you can find more than 25,000 Nordic Swan ecolabelled products. 93% of all Nordic consumers recognise the Nordic Swan Ecolabel as a brand, and 74% believe that the Nordic Swan Ecolabel makes it easier for them to make envi¬ronmentally friendly choices (IPSOS 2019).

Denmark, Finland, Norway, Iceland, Sweden

Securing a sustainable future
The Nordic Swan Ecolabel works to reduce the overall environmental impact from production and consumption and contributes significantly to UN Sustainable Development Goal 12: Responsible consumption and production.
To ensure maximum environmental impact, the Nordic Swan Ecolabel sets product specific requirements and evaluates the environmental impact of a product in all relevant stages of a product lifecycle - from raw materials, production, and use, to waste, re-use and recycling.
Common to all products certified with the Nordic Swan Ecolabel is that they meet strict environmental and health requirements. All requirements must be documented and are verified by Nordic Ecolabelling. Nordic Ecolabelling regularly reviews and tightens the requirements.
Therefore, certifications are time-limited and companies must re-apply to ensure sustainable development.

International website:
Nordic-ecolabel.org
National websites:
Denmark: ecolabel.dk
Sweden: svanen.se
Norway: svanemerket.no (in Norwegian)
Finland: joutsenmerkki.fi (in Finnish)
Iceland: svanurinn.is (in Icelandic)

Thailand

The Thai Green Label Scheme was initiated by the Thailand Business Council for Sustainable Development (TBCSD) in October 1993. It was formally launched in August ১৭৭৪ by The Thailand Environment Institute (TEI) and Thai Industrial Standards Institute (TISI). The Green Label is an environmental certification logo awarded to specific products which have less detrimental impact on the environment in comparison with other products serving the same function. The Thai Green Label Scheme applies to all products and services, but not foods, beverage, and pharmaceuticals. Products or services which meet the Thai Green Label criteria may carry the Thai Green Label. Participation in the scheme is voluntary.

Thailand Environment Institute (TEI)
16/151 Muang Thong Thani, Bond Street,
Bangpood, Pakkred, Nonthaburi 11120 THAILAND
Tel. +66 2 503 3333 ext. 303, 315, 116
Fax. +66 2 504 4826-8
Website: http://www.tei.or.th/greenlabel/
Email: lunchakorn@tei.or.th

EUROPE

Established in 1992 and recognized across Europe and worldwide, the EU Ecolabel is a label of environmental excellence that is awarded to products and services meeting high environmental standards throughout their life-cycle: from raw material extraction, to production, distribution and disposal. The EU Ecolabel promotes the circular economy by encouraging producers to generate less waste and CO_2 during the manufacturing process. The EU Ecolabel criteria also encourages companies to develop products that are durable, easy to repair and recycle.

The EU Ecolabel criteria provide exigent guidelines for companies looking to lower their environmental impact and guarantee the efficiency of their environmental actions through third party controls. Furthermore, many companies turn to the EU Ecolabel criteria for guidance on eco-friendly best practices when developing their product lines. The EU Ecolabel helps you identify products and services that have a reduced environmental impact throughout their life cycle, from the extraction of raw material through to production, use and disposal. Recognised throughout Europe, EU Ecolabel is a voluntary label promoting environmental excellence which can be trusted.

Spain , Germany, Italy, Sweden, Greece, Portugal, Poland, Belgium, Netherlands, Estonia, Finland, Austria, Lithuania, Czech Republic, Norway, Cyprus, Ireland, Slovenia, Hungary, Romania, Croatia, Bulgaria, Malta, Slovak Republic, Latvia, Luxembourg, Iceland

Contact and more information via: http://ec.europe.eu

CHAPTER 5

Type II Environmental Labelling

Type II environmental labelling refers to the claims made on product labels in connection with business centers. This includes familiar claims such as recyclable, ozone-free, 60% phosphate-free, and the like. This type of labelling can be in the form of a mark or sentence on the product packaging. Some of them are valid environmental claims—and some can be completely misleading.

Usually, all countries have laws against deceptive advertisements, so why has the International Organization for Standardization discussed this issue? The answer is that it is not clear whether the environmental claims have a technical basis or whether the ad is meaningless.

Most countries have guidelines at the national level to help producers and consumers know what constitutes a true, scientifically valid claim.
There is a national standard on this in Canada. In Australia, the Consumer Commission has published guidance on this, and there are similar examples in other countries.

GREEN BICYCLE
your best vehicle

BE RESPONSIBLE
protect the forest

TIME FOR CHANGE
stop global warming

100% ECO
natural product

BE ECO FRIENDLY
plant your own tree

PROTECT WILDLIFE
save the animals

PLANT AN ACORN
we need more trees

ECO FRIENDLY
window cleaner

SAVE THE WATER
every drop counts

ECO LESSONS
learn how to live

ECO AGREEMENT
your green contract

ECO HOUSEHOLD
your green home

ECO DROP
your reusable water

ECO AQUA
your reusable bottle

ECO TAP
your natural stream

RECYCLING DAY
save the nature

ECO COUNTER
use minimal

ECO WATER
your green choice

ECO RAIN
your free water

ECO AGRICULTURE
your rich harvest

ECO BARREL
your green reservoir

ECO FOUNTAIN
your traditional way

ECO SOIL WATER
your pump fountain

ECO CAR
clean environment

ECO WATERS
your clean resources

ECO DECOR
a drop of green

ECO FRIENDLY
dishwashing liquid

LIME JUICE
your natural cleaner

ECO PLANTS
your green medicine

ECO LEAF
your green choice

ECO FRIENDLY
floor cleaner

ECO BATHROOM
without chemicals

ECO NATURE
your fresh air

ECO GARDEN
productive soil

WHITE VINEGAR
your natural cleaner

ECO SPONGE
your perfect cleaner

Canada

Environmental Sustain for Future kids established in Vancouver, BC Canada in 2020. (ESFK) is an international ecolabel focused on taking care of environment for future of kids.

ESFK defined as 'self-declared' environmental claims made by manufacturers and businesses based on ISO 14020 series of standards, the claimant can declare the environmental objectives and targets in relation to taking care of environment for future kids. However, this declaration will be verifiable.

Environmental Sustain for Future Kids
Vancouver, BC CANADA

Email: info@esfk.org
Web: www.esfk.org

Eco-Friendly Homemade Laundry Detergent

CHAPTER 6

Type III Environmental Labelling

Type III environmental labelling is a distinct form of third-party environmental labelling pattern designed to avoid the difficulties that can result from type I labelling. Technical committee for Environment of International organization for Standardization has undertaken a new project to standardize guidelines and Type III labelling methods. One of the main objections raised by industries to Type I labelling is the basis for its management.

Due to the nature of the system, less than 50% of the various products on the market can meet the criteria and qualify for Type I Labelling. As long as the industry is the main supporter of other third-party models for quality systems, it is sometimes difficult for an industry to support a program that can only benefit 15% of its members. This type of labelling is currently practiced in some countries, such as Sweden, Canada, and the United States. Choosing the right product has never been easy, but Type III labelling will help because each product can have a label that describes its environmental performance and is certified by a third-party company. Consumers can then compare labels and choose their favorite products.

CHAPTER 7

All about 'Eco-friendly' Cleaning Products

Green cleaning products should not contain hazardous chemicals, and so they are likely to pose fewer health risks. Green cleaning products are less hazardous for the environment, too. They do not contain chemicals that cause significant air or water pollution and are often in recyclable or recycled packaging. The eco-friendly cleaning products, also known as green cleaning products, are made from plant-based ingredients, natural colors or fragrances, uses eco-friendly packaging methods, and are biodegradable. Support sustainable human and ecological use and reuse of remediated land; Minimize impacts to water quality and water resources; Reduce air toxics emissions and greenhouse gas production; Minimize material use and waste production; and Conserve natural resources and energy.

How do I know which cleaning products are the most environmentally friendly?

Almost all people from all over the world use household cleaning products from dish detergents to bathroom cleaners and floor polish to scouring pads. Most of us are exposed to cleaners on a daily basis,... Even if we don't use cleaners, it's likely we're regularly come into contact with them at work, school or elsewhere.

Unfortunately, cleaners often contain harsh chemicals that can be harmful to our health and planet. Health effects associated with cleaning products include asthma, contact dermatitis, burns to the skin and eyes and inflammation or fluid in the lungs. Long-term repercussions may include reproductive problems, cancer, heart disease and other health issues. The environment also can fall victim to cleaning products' acrid ingredients. Chemicals in laundry detergents, for example, have been found in 75 percent of streams

and waterways throughout in different countries. Some ingredients in cleaners have been directly linked to environmental problems, such as chemicals getting into bodies of water and foaming in streams, and some commonly used household cleaner ingredients have room for improvement even today. Health and environmental concerns have prompted many consumers to push for safer alternatives to cleaning products. **But identifying environmentally safe cleaners can be challenging for consumers.** With so many product options, choosing the safest, healthiest cleaners for the home can be challenging for reasons other than too many choices, namely the lack of a national regulatory body. A fraction of the tens of thousands of chemicals in commerce in different countries are used in consumer goods like household cleaners. Chemicals are regulated as they enter commerce rather than at the product level.

But shoppers who want to know exact ingredients might not find what they're looking for on household cleaner labels. In many countries law does not require manufacturers of cleaning products to list all ingredients on labels. But manufacturers might be changing their ways in the near future. Some of the associations, launched a joint, voluntary effort to encourage their members to list their ingredients in a public format. In the meantime, consumers are left to make sense of what's on the packaging.

Different manufacturers can make the same marketing claim like "degradable" or "ozone-friendly" and mean different things with those terms. This has resulted in confusion among consumers.

Besides looking for the Design for Environment Labellings and knowing what marketing terms mean, consumers can also read product packaging to make sure environmental claims are qualified. All assertions should specify whether it's referring to the product, packaging or both. Similarly, Environmental labels should come with an explanation and identify the third party doing the certifying.

The organization should be independent from advertisers and have expertise in the area for which it's certifying. Other indicators of environmental responsibility are the following: recycled, recyclable or refillable containers; concentrated products that require less packaging; cleaners free of chlorofluorocarbons (CFCs) that can deplete the ozone; and degradable, biodegradable or photodegradable product contents or packaging.

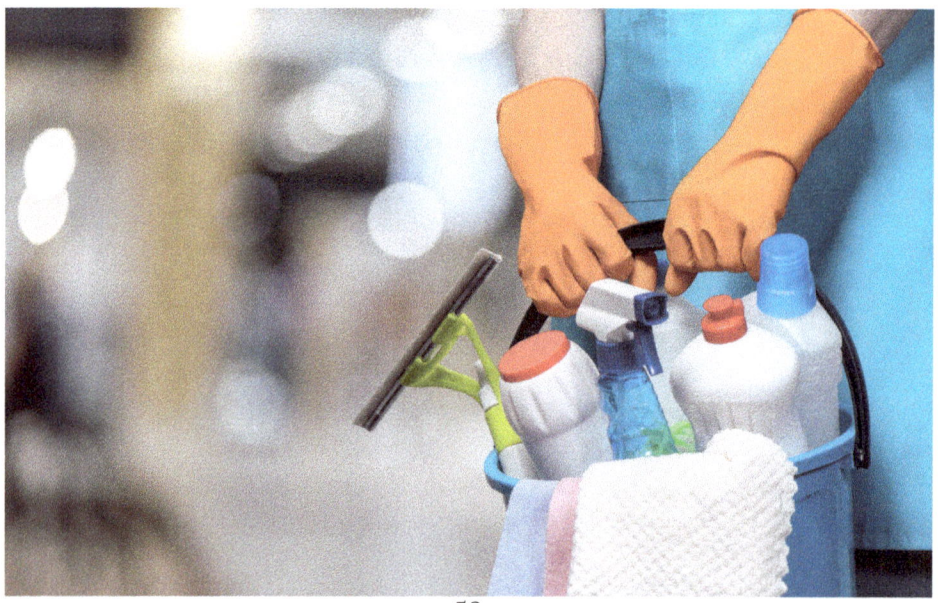

For example, a toilet cleaner ad that claims the solid waste generated by disposing of its container is "now 20 percent less than our previous container," is in good practice if the cleaner company can prove disposal of the new package contributes 20 percent less waste by weight or volume to the solid waste stream. Comparatively, the general claim "20 percent less waste" is ambiguous and therefore deceptive because it's unclear if the claim is referring to a preceding product or that of a competitor, according to the general principles on International Environmental labelling (page: 20-21).

Homemade Cleaning Products: Natural, Green, Eco-Friendly

A mixture of vinegar and baking soda can do wonders for your cleaning needs. This combination can be used in many ways to fight against severe stains, so you do not need to run out to the grocery store to buy a solution filled with chemicals anymore. Not only will natural cleaners make your life better, they will virtually eliminate that bad smell in the house and they're surprisingly inexpensive to create.

Chemical - free Recipes for Homemade Cleaning Products:

If you're wanting to pitch those toxic, commercial household cleaners and switch to natural, homemade cleaner, these simple recipes will have you cleaning green in no time, Before we get to the cleaning, let's check out some of the most common (and most useful) non-toxic cleaning products:

Baking Soda
Baking soda is a pantry staple with proven virus-killing abilities that also effectively cleans, deodorizes, brightens, and cuts through grease and grimeTrusted Source.

Castile Soap
Castile soap is a style of soap that's made from 100 percent plant oils (meaning it uses no animal products or chemical detergents).

Vinegar
Thanks to its acidity, vinegar is nothing short of a cleaning wunderkind—it effectively (and gently!) eliminates grease, soap scum, and grime.

Lemon Juice
Natural lemon juice annihilates mildew and mold, cuts through grease, and shines hard surfaces (It also smells awesome.).

Olive Oil
This good-for-you cooking oil also works as a cleaner and polisher.

Essential Oils
Essential oils have gained popularity thanks to aromatherapy, but these naturally occurring plant compounds also make great scent additions to homemade cleaning products (particularly if you're not into the smell of vinegar). Essential oils are generally considered safe, but these extracts can trigger allergies—so keep this in mind when choosing scents.

Borax

Many DIY cleaners tout Borax (a boron mineral and salt) as a non-toxic alternative to mainstream cleaning products; however, the issue is pretty hotly debated. Some research suggests Borax can act as a skin and eye irritant and that it disrupts hormones. For this list, we've chosen to avoid products that use Borax.

A note on mixing products:

Most of these ingredients can be used in combination with each other; however, many sources advise against mixing castile soap with vinegar or lemon juice. Since castile soap is basic (i.e., high on the pH scale) and vinegar and lemons are acidic, the products basically cancel each other out when used in combination (though it's fine to wash with a base—like castile soap—and rinse with an acid—like vinegar!).

Cleaning Recipes

Many of these cleaners can be used in multiple places, but we've assigned them to particular areas for easy reference:

- **Bathroom**
- **Kitchen**
- **Lundry Room**
- **Others**

Bathroom

1. Toilets

For a heavy-duty toilet scrub that deodorizes while it cleans, pour ½ cup of baking soda and about 10 drops of tea tree essential oil into the toilet. Add ¼ cup of vinegar to the bowl and scrub away while the mixture fizzes.

For daily cleaning, fill a small spray bottle with vinegar (about 1 cup should do it) and a few drops of an essential oil of your choosing (lemon and tea tree both work well). Spray on the toilet seats, let it sit for a few minutes, and then wipe the surface clean.

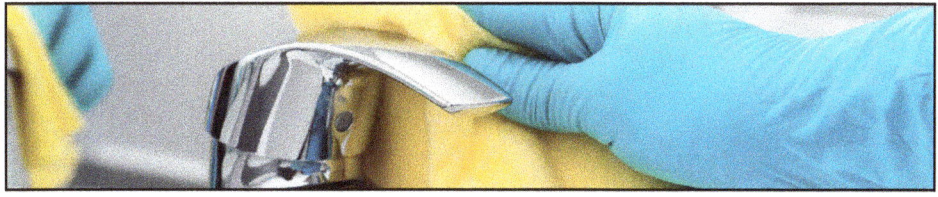

2. Tub and Shower

Tubs and showers can produce some of the toughest grime, but it's no match for the cleaning power of vinegar. To get rid of mildew, spray pure white vinegar on the offending area, let it sit for at least 30 minutes, and then rinse with warm water (don't be afraid to use a sponge if rinsing doesn't clear away the grossness on its own). Alternatively, try mixing together baking soda with a bit of liquid castile soap, then scrub and rinse.

For daily cleaning or to get rid of soap scum, mix 1 part water with 1 part vinegar (and a few drops of essential oils if you're not into the smell of vinegar) in a spray bottle. Spray, let it sit for at least several minutes, and then wipe away.

3. Disinfectant

Skip the bleach and make a homemade germ-killer instead. Just mix 2 cups of water, 3 tablespoons of liquid soap, and 20-30 drops of tea tree oil. Voila!

4. Air Freshener

Defeat less-visible bathroom "uncleanliness" with this homemade, non-toxic air freshener. All you need is baking soda, your favorite essential oil, and an old jar with a lid you don't mind poking holes in (follow the link for full instructions).

5. Hand Soap

Once you're done cleaning the bathroom, it's time to make yourself clean (or at least your hands). To make a non-toxic, foaming hand soap, mix together liquid castile soap and water (and an essential oil if you feel like it) in a foaming soap dispenser. Fill about one fifth of the bottle with soap, then top it off with water.

Kitchen

6-Countertops
For a simple, all-purpose counter cleaner, mix together equal parts vinegar and water in a spray bottle. If your countertop is made from marble, granite, or stone, skip the vinegar (its acidity is no good for these surfaces) and use rubbing alcohol or the wondrous power of vodka instead.

7. Cutting Boards
Talk about non-toxic: All that's needed to clean and sanitize cutting boards (wood or plastic) is… a lemon! Cut it in half, run it over the surfaces, let sit for ten minutes, and then rinse away. If you need some serious scrubbing power, sprinkle some coarse or Kosher salt over the board, and then rub with ½ a lemon.

8. Oven
To clean stubborn, caked-on food out of the oven, just heat the over to 125 degrees and grab your spray bottle of vinegar (see "countertops" above). Once the oven is warm, spray the caked-on stuff until it's lightly damp and then pour salt directly onto the affected areas. Turn off the oven, let it cool, and then use a wet towel to scrub away at the mess. If that doesn't cut it, follow the same instructions but try use baking soda in place of salt (just let it sit for a few minutes before scrubbing).

9. Garbage Disposal

This one is so cool. Pour 1 cup of vinegar into an ice cube tray and top off the slots with water. Once they're frozen, toss a few down the disposal and let it run—doing so should remove any food that was stuck to the blades.

10. Microwave

It's easy to overlook the microwave while cleaning, but man can it get gross in there. To combat the gunk, pour some vinegar into a small cup and mix in a little lemon juice (exact amounts don't really matter). Put the cup in the microwave, let the microwave run for 2 minutes, and leave the door closed for several more minutes. Finally, open the door and simply wipe down all the sides with a warm cloth or sponge—no scrubbing required!

11. Sink Drain

To unclog a stuffed-up drain, start by boiling about 2 cups of water. Pour ½ cup of baking soda into the drain, and then add the water while it's still nice and hot. If that doesn't do the trick, follow the baking soda with ½ cup of vinegar, cover it up tightly (a pot lid should work nicely), wait until the fizzing slows down (when baking soda and vinegar come in contact, they'll react by fizzing) and then add one gallon of boiling water.

12. Pan De-Greaser

To cut through the grime on frying pans, simply apply some salt (no water necessary) and scrub vigorously.

13. Cast-Iron Pans

Kitchen professionals are pretty against using soap, steel wool, or dishwashers to clean cast-iron pans. Luckily, there's an alternative way to tackle cast-iron grossness: combine olive oil and a teaspoon of coarse salt in the pan. Scrub with a stiff brush, rinse with hot water, and you're done!

14. Dishwasher Detergent

If you're lucky enough to have a dishwasher, simply mix together 1 cup of liquid castile soap and 1 cup of water (2 teaspoons of lemon juice optional) in a quart-size glass jar. Add some of this mixture to one detergent compartment of the dishwasher, and fill the other compartment with white vinegar.

15. Dish Soap

If washing dishes by hand, simply combine 1 cup of liquid castile soap and 3 tablespoons water (a few drops of essential oil optional) in a bottle of your choice. Shake well and use like you would any other dish soap.

16. Refrigerator Cleaner

To clean what is perhaps the toughest of all kitchen "gross spots," reach for the baking soda. Add about ½ cup of the white stuff to a bucket of hot water. Dip a clean rag in the mixture and use it to wipe down the fridge's insides.

17. Bleach

For serious disinfectant power, mix ½ cup baking soda, 1 teaspoon castile soap, and ½ teaspoon hydrogen peroxide. Use a cloth to apply the mixture to a wet surface, scrub, and then rinse thoroughly.

Lundry Room

18. Laundry Detergent
It's tough to come by homemade laundry detergents that don't use Borax, but give this one a try. The recipe calls for glycerin soap, washing soda, baking soda, citric acid, and coarse salt. For full instructions, follow the link!

19. Fabric Softener
Skip the liquid fabric softener and make clothes nice and snuggly the non-toxic way. Make a big batch of softener by adding 20-30 drops of the essential oil of your choice to a one-gallon jug of white vinegar. Add 1/3 cup to each laundry load (just be sure to shake the mixture prior to each use).

20. Laundry "Scenter"
To add a fresh, clean scent to laundry, make a sachet stuffed with your favorite dried herbs (lavender, peppermint, and lemon verbena are all great options). Toss it in the dryer while it's in use, and voila: customized, non-toxic scent!

21. Bleach
For a nontoxic laundry bleach alternative, add some lemon juice to the rinse cycle.

Everything Else

22. Floors
For a simple, effective tile floor cleaner, simply combine one part white vinegar with two parts warm water in a bucket. Use a mop or rag to scrub down the floors with the solution. No need to rinse off! (Note: this one's not recommended for wood floors).

23. Walls
To scrub down walls, mix ¼ cup white vinegar with 1 quart warm water, then use a rag to scrub those walls down. To remove black marks, simply scrub at the spot with a little bit of baking soda.

24. Windows and Mirrors
For an all-purpose window cleaner, combine 1 part white vinegar with 4 parts water (feel free to add some lemon juice if you're feeling citrusy), then use a sponge or rag to scrub away.

25. Furniture Polish
For an all-purpose furniture polish, combine ¼ cup vinegar with ¾ cup olive oil and use a soft cloth to distribute the mixture over furniture. For wood furniture (or as an alternative to the first recipe), combine ¼ cup lemon juice with ½ cup olive oil, then follow the same procedure.

26. Silver Cleaner

Put silver utensils and jewelry back to good use the non-toxic way. Line a sink or bucket with aluminum foil, lay out the silver on top of the aluminum, and pour in boiling water, 1 cup of baking soda, and a pinch of salt. Let it sit for several minutes and watch as—like magic—the tarnish disappears! Note: If you're concerned about immersing a particular item, simply rub it with toothpaste and a soft cloth, rinse it with warm water, and allow it air to dry.

27. Wood Cleaner

Clean varnished wood by combining 2 tablespoons of olive oil, 1 tablespoon of white vinegar, and a quart of warm water in a spray bottle. Spray onto wood and then dry with a soft cloth. (Note: Since olive oil can leave behind some (slippery) residue, this one might not be the best option for wood floors.)

Important Note: We've done our absolute best to provide the best information possible, but since we haven't tried every single one of these solutions in every possible cleaning situation, we can't vouch for them 100 percent.

What's the difference between products that disinfect, sanitize, and clean surfaces?

Products used to kill viruses and bacteria on surfaces are considered as antimicrobial pesticides. Sanitizers and disinfectants are two types of antimicrobial pesticides.

Cleaning	Cleaning removes dirt and organic matter from surfaces using soap or detergents.
Sanitizing	Sanitizing kills bacteria on surfaces using chemicals. It is not intended to kill viruses.
Disinfecting	Disinfecting kills viruses and bacteria on surfaces using chemicals.

Experts agree that frequent handwashing is one of the first lines of defense against many illnesses. But no matter how many times you wash your hands, there are always some sneaky little germs lurking around to hitch a ride on your skin. They loiter on shopping cart handles, linger on light switches, lurk about the phone and even hang around on the remote controls. That's why disinfectants and disinfecting cleaners can be a helpful option.

Why Disinfect
- Regular cleaning products do a good job of removing soil and many germs. Disinfectants or disinfectant cleaners are able to go further and kill many of those germs.
- Surfaces may be contaminated even when they're not visibly soiled.
- Germs can live on surfaces for hours or even days.

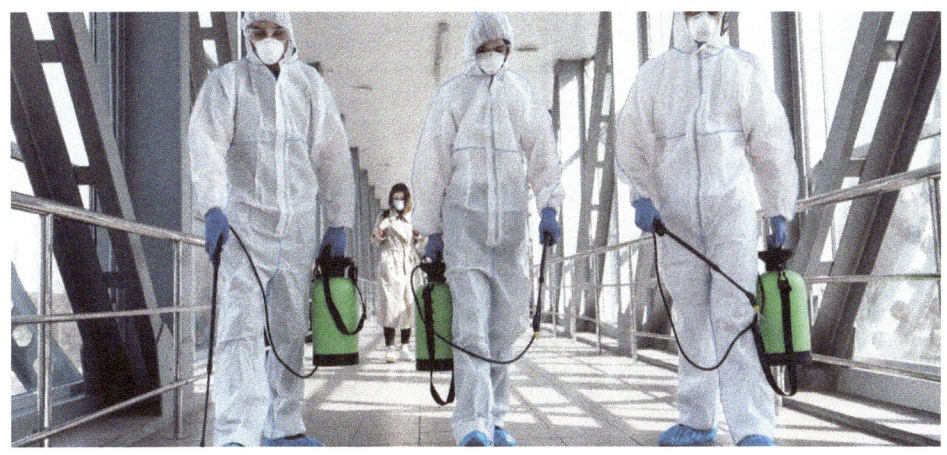

How to Disinfect
- Read the label before using any cleaning or disinfecting product to ensure you are following the directions for use and storage instructions.
- Pre-clean any surfaces prior to disinfecting to remove any excess dirt or grime.
- Apply the disinfectant, then the surface needs to stay wet for the entire time indicated on the product label; this is called contact time.
- If disinfecting food contact surfaces or toys, rinse with water after they air dry.
- When disinfecting, target surfaces that are frequently touched, especially if someone in the home is ill.
- If using a disinfectant wipe, throw out after using. Do not flush any non-flushable products.

What to Look for in a Disinfectant

Products that say "Disinfectant" on the label are required to meet legal specifications. To be sure the product has met all defined requirements for effectiveness, look for Registration Number on the label. You must follow the product label instructions exactly for the disinfectant to be effective. Your choices include:

Chlorine bleach. It disinfects when mixed and used properly. Read the label for instructions.

Disinfectant cleaners. These dual purpose products contain ingredients that help remove soil as well as kill germs.

Disinfectants. These products are designed to be effective against the germs indicated on their labels. Surfaces should be clean prior to disinfecting.

Some of the more frequently used active ingredients are sodium hypochlorite, ethanol, pine oil, hydrogen peroxide, citric acid and quats (quaternary ammonium compounds).

When killing surface germs is your goal, look for products that are disinfectants, some common disinfectants include Quaternary Ammonium Compounds (QACs), commonly referred to as Quats.

The Future of Cleaning

Cleaning technology has come a long way from the ancient Babylonian way of soap-making. Today's cleaning products are the result of thoughtful design, experimentation, and safety testing.

The machines we use to clean have also improved, becoming more sustainable and friendly for our environment. So far we have been able to make new cleaning products that allow us to wash in cold water (saving energy from water heating), wash with less water, and make packaging smaller (to save material and avoid shipping extra weight).

Future scientists will have a great opportunity to continue to create new cleaning design products that will continue to keep us healthy and do even more to help protect human health and the environment.

Nanotechnology is the future, not only for things like computers and building materials but also for cleaning products. Not only is it important to focus on keeping surface areas clean, but we should also keep in mind air quality.

Many workers have transitioned to more remote work over the past year, meaning more time spent inside your home with who knows what kind of germs or contaminants lingering. Some air-purification systems don't always catch pollutants at room temperature.

Nanotechnology aids in the filtration of such pollutants. Air-purifier sprays like Purbloc's NANO GRAB are your ticket to a clean, germ-free, home.

Harmful toxins leached out by ordinary household cleaners can cause everything from chapped hands to potential lung issues in the long term. Long-term health issues can then translate to high costs.

By going with a natural cleaner you're thinking of not only your long-term health, but you're saving money in the long run by taking care of yourself now and those around you with safe and effective supplies.

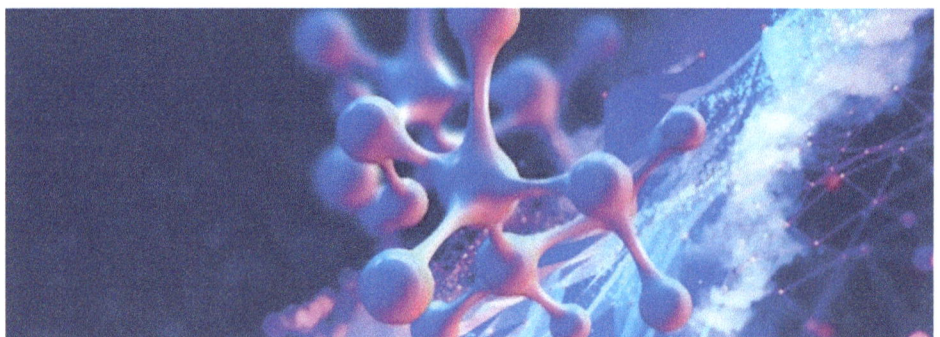

Keep in Mind

There are so many incredible real-world benefits of nanotechnology. A summation of a few products and processes enhanced by such benefits:

- Coatings to reduce cleaning efforts
- Improved energy efficiency
- Fabrics with increased resistance to stains
- Extra durable goods that increase the life of a product thus creating less waste in the long run
- Easier and more efficient water filtration
- Insulation materials reducing the energy needed to heat and cool buildings

Families and businesses alike can all benefit from the efficient uses mentioned above. From the reduction of energy consumption to the wonders of regenerative medicine, nanotechnology is a key player in the race toward enhancing and caring for our way of living.

Chilliwack, British Columbia, Canada

Consumers are increasingly demanding sustainable products that are not toxic to themselves or the environment. The natural market is growing exponentially, and choosing raw, natural materials will cement your brand as a safe choice — both environmentally and economically

CHAPTER 9

Top Ten Award International Network Environmental Pioneers

Top Ten Award international Network (TTAIN) was established in 2012 to recognize outstanding individuals, groups, companies, organizations representing the best in the public works profession. TTAIN publishing books related to international Eco-labeling plans to increase public knowledge in purchasing based on the environmental impacts of products. We introduce in each volume some of the organizations that are doing their best in relation to taking care of the environmnet.

Canada

ECOCERT Canada, a subsidiary of the ECOCERT GROUP, has and continues to assist stakeholders in the implementation and promotion of sustainable practices through certification, consulting and training services.

Committed for over 25 years to organic agriculture with Garantie Bio, ECOCERT has become the benchmark for organic certification in Canada. With more than 100 employees and its 4 offices across the country, ECOCERT Canada offers first-rate, client-focused service in both French and English to many sectors. Certification against the Ecological cleaning products will enable you to label your product as "natural" or "organic." All the products marketed with the ECOCERT logo have been verified by our teams: from composition to processing and packaging.

Consumers are thus given transparent information on the content of natural and organic ingredients that are listed on the product label. Requirements for labeling: Natural origin All the ingredients are derived from natural origin except those included in a restrictive approved ingredients list (including preservatives) which are authorized in small quantities.

Why choose Ecocert?

With nearly 30 years of experience for audit and certification of organic products in France and in more than 130 countries, Ecocert is the world's leading specialist in the certification of sustainable practices.

Contact:
Web: https://www.ecocert.com/en-CA/
Tel.: (+1) 418-838-6941

Note:
We've done our absolute best to provide the best information possible, but since we haven't tried every single one of these solutions in every possible cleaning situation, we can't vouch for them 100 percent.

UNEP

The United Nations Environment Programme (UNEP) is the leading global environmental authority that sets the global environmental agenda, promotes the coherent implementation of the environmental dimension of sustainable development within the United Nations system, and serves as an authoritative advocate for the global environment.

Our mission is to provide leadership and encourage partnership in caring for the environment by inspiring, informing, and enabling nations and peoples to improve their quality of life without compromising that of future generations.

Headquartered in Nairobi, Kenya, we work through our divisions as well as our regional, liaison and out-posted offices and a growing network of collaborating centres of excellence. We also host several environmental conventions, secretariats and inter-agency coordinating bodies. UN Environment is led by our Executive Director.

We categorize our work into seven broad thematic areas: climate change, disasters and conflicts, ecosystem management, environmental governance, chemicals and waste, resource efficiency, and environment under review. In all of our work, we maintain our overarching commitment to sustainability.

Website: www.unep.org

Bibliography

Bibliography:

Amberg, N.; Magda, R. Environmental Pollution and Sustainability or the Impact of the Environmentally Conscious Measures of International Cosmetic Companies on Purchasing Organic Cosmetics. Visegrad J. Bioecon. Sustain. Dev. 2018, 1, 23.

Asadi, J., "International Environmental Labelling, Economic Consequencies, Export Magazine, July 2001

Asadi, J. 2008. Mobile Phone as management systems tools, ISO Magazine, Vol.8, No.1

Asadi, J., Eco-Labelling Standards, National Standard Magazine, Sep. 2004.

Barbieux, D.; Padula, A.D. Paths and Challenges of New Technologies: The Case of Nanotechnology-Based Cosmetics Development in Brazil. Adm. Sci. 2018, 8, 16.

Basketter, D.; Corsini, E. Can We Make Cosmetic Contact Allergy History? Cosmetics 2016, 3, 11.

Benitta Christy P & Dr. Kavitha S, "GO-GREEN TEXTILES FOR ENVIRONMENT", Advanced Engineering and Applied Sciences: An International Journal 2014; 4(3): 26-28

Chemical Week, 1999. Europe's Beef Ban Tests Precautionary Principle. (August 11).

Chaudri, S.K.; Jain, N.K. History of Cosmetics. Asian J. Pharm. 2009, 7–9, 164–167.

CHOI, J.P. Brand Extension as Informational Leverage. Review of Eco- nomic Studies, Vol. 65 (1998), pp. 655-669.

Conway, G. 2000. Genetically modified crops: risks and promise.

Corrado, M., (1989), The Greening Consumer in Britain, MORI, London

Corrado, M., (1997), Green Behaviour – Sustainable Trends, Sustainable Lives?, MORI, london, accessed via countries. Manila, Asian Development Bank 33p.

Cosmetics, Perfume, & Hygiene in Ancient Egypt. Available online: https://www.ancient.eu/article/1061/cosmetics-perfume--hygiene-in-ancient-egypt/ (accessed on 4 May 2017).

Davies, Clive. Chief, Design for the Environment Program, Environmental Protection Agency. Interview. March 24, 2009.

Federal Trade Commission, "Sorting Out Green Advertising Claims." http://www.ftc.gov/bcp/edu/pubs/consumer/general/gen02.shtm (March 26, 2009, March 27, 2009)

MSNBC, "Do You Know What's in Your Cleaning Products?" http://today.msnbc.msn.com/id/29663739/ (March 17, 2009)

Ooyen, Carla. Research Manager with Nutrition Business Journal. Personal correspondence. March 19, 2009.

Tekin, Jenn. Marketing Manager with Packaged Facts & SBI. Personal correspondence. March 17, 2009.

University of California - Berkeley. http://berkeley.edu/news/media/releases/2006/05/22_householdchemicals.shtml (March 26, 2009)

U.S. Department of Health and Human Services, Household Products Database.http://householdproducts.nlm.nih.gov/cgi-bin/household/prodtree?prodcat=Inside+the+Home (March 17,

Women's Voices of the Earth, "Household Cleaning Products and Effects on Human Health."http://www.womenandenvironment.org/campaignsandprograms/SafeCleaning/safecleaninghealth (March 17, 2009)

EMONS, W. Credence Goods Monopolists. International Journal of In- dustrial Organization, Vol. 19 (2001), pp. 375-389.

European Union official website: https://ec.europa.eu/info/about-european-commission/contact_en

Feenstra, R.C. "Exact Hedonic Price Indexes," Review of Economics and Statistics 77 (1995): 634-653.

Feenstra, R.C., and J.A. Levinsohn. "Estimating Markups and Market Conduct with Multidimensional Product Attributes," Review of Economic Studies (62 (1995): 19-52.

Forest Stewardship Council: "Principles and criteria for forest stewardship" Document 1.2: <http://www.fscoax.org>

Forsyth, K. 1999. Will consumers pay more for certified wood products? Journal of Forestry 97 (2) : 18-22.

Freeman, A. M III. The Measurement of Environmental and Resource Values. Theory and Methods. Washington D.C.: Resource for the Future, 1993.

Friends of the Earth, 1993. Timber certification and eco-labeling. London, FOE:

Geetha Margret Soundri, "Ecofriendly Antimicrobial Finishing of Textiles Using Natural Extract", Journal of International Academic Research For Multidisciplinary, ISSN: 2320 – 5083, 2014, Vol 2.

Graves, P., J.C. Murdoch, M.A. Thayer, and D. Waldman. "The Robustness of Hedonic Price Estimation: Urban Air Quality," Land Economics 64(1988): 220-233.

Halvorsen, R. and R. Palmquist. "The Interpretation of Dummy Variables in Semilogarithmic Equations." American Economic Review 70:474-75 (1980).

Imhoff, Dan, and Grose, Lynda, and Carra, Roberto., "Organic Cotton Exhibit," Mimeo. Simple Life and distributed the Texas Organic Cotton Marketing Cooperative, O'Donnell, Texas (1996).

Imhoff, Dan. "Growing Pains: Organic Cotton Tests the Fibre of Growers and Manufacturers Alike," reprinted on Simple Life's web page (simplelife.com), but first printed by Farmer to Farmer, December 1995.

Incomplete Consumer Information in Laboratory Markets. Journal of Environmental labeling.

ISO 14020, ISO 14021,ISO 14024,ISO 14025, International Organization for Standardization.

Kennedy, P.E. "Estimation with Correctly Interpreted Dummy Variables in Semilogarithmic Equations," American Economic Review 71: 801 (1981).

Kirchho®, S., (2000), Green Business and Blue Angels.

Kraus, Jeff. Lab Technician at the North Carolina School of Textiles.

Labeling Issues, Policies and Practices Worldwide.

Lamport, L. 1998. The cast of (timber) certifiers: who are they? International J. Ecoforestry 11(4): 118-122.

Large Scale impoverishment of Amazonian forests by logging and fire. 1999.

Lathrop, K.W. and Centner, T.J. 1998. Eco-labeling and ISO 14000: An analysis of US regulatory systems and issues concerning adoption of type II standards. Environmental

Lee, J. et al. 1996. Trade related environmental measures; sizing and comparing impacts.

Lehtonen, Markku. 1997. Criteria in Environmental Labeling: A comparative Analysis on Environmental Criteria in Selected Labeling Schemes. Geneva, UNEP. 148p.

LIEBI, T. Trusting Labels: A Matter of Numbers? Working Paper Uni versity of Bern, No. 0201 (2002).

Lindstrom, T. 1999. Forest Certification: The View from Europe's NIPFs. Journal of Forestry 97(3): 25-31. London

Losey, J.E., Rayor, L.S. & Carter, M.E. 1999. Transgenic pollen harms monarch larvae. Nature 399 20 May): p.214.

Management 22 (2) : 163-172.

Mattoo, A. and H. V. Singh, (1994), Eco-Labelling: Policy Considera-Michaels, R. G., and V. K. Smith. "Market Segmentation And Valuing Amenities With Hedonic Models: The Case Of Hazardous Waste Sites," Journal of Urban Economics, 1990 28(2), 223-242.

Nicholson-Lord, D., (1993) 'Tis the Season to be Green, The Independent, 20 December

Nuttall, N., (1993), Shoppers can cross green products off their lists, The Times, 3 July

OCDE/GD(97)105. Paris, OECD. 81p.

OECD. "Ec-labelling: Actual Effects of Selected Programmes," OCDE/GD (97) 105, 1997, Paris. (available on line at http://www.oecd.org/env/eco/books.htm#trademono)

OECD. 1997a. Case study on eco-labeling schemes. Paris, OECD (30 Dec):

OECD. 1997b. Eco-labeling: Actual Effects of Selected Programs.

Osborne, L. "Market Structure, Hedonic Models, and the Valuation of Environmental Amenities." Unpublished Ph.D. dissertation. North Carolina State University, 1995.

Osborne, L., and V. K. Smith. "Environmental Amenities, Product Differentiation, and market Power," Mimeo, 1997.

Ozanne, L.K. and Vlosky, R.P. 1996. Wood products environmental certification: the United States perspective". Forestry Chronicle 72 (2) : 157-165.

Palmquist, R. B., F. M. Roka, and T.Vukina. "Hog Operations, Environmental Effects, and Residential Property Values," Land Economics 73(1), (1997): 114-24.

Palmquist, R.B. "Hedonic Methods," in J.B Braden and C.D. Kolstad, eds. Measuring the Demand for Environmental Improvement. Amsterdam, NL: Elsevier, 1991.

Pento, T. 1997. Implementation of Public Green Procurement Programs (22-31) in Greener Purchasing: Opportunities and Innovations. Sheffield, Greenleaf Publ. 325 p.

Perloff, J. "Industrial Organization Lecture Notes," Mimeo. University of California at Berkeley (1985).

Plant, C. and Plant, J. 1991. Green business: hope or hoax? Philadelphia, New Society Publishers 136 p.

Polak, J. and Bergholm, K. 1997. Eco-labeling and trade: a cooperative approach (Jan.): Policy in a Green Market. Environmental and Resource Economics 22, 419-

Poore, M.E.D. et al. 1989. No timber without trees. London, Earthscan. 352p.

Raff, D. M.G., and M. Trajtenberg. "Quality-Adjusted Prices for the American Automobile Industry: 1906-1940." NBER Working Paper Series, Working Paper No. 5035, February 1995.

Roberts, J. T. 1998. Emerging global environment standards: prospects and perils. Journal of Developing Societies 14 (1): 144-163.

Rosen, S., "Hedonic Prices and Implicit Markets: Product Differentiation in Pure Competition." Journal of Political Economy. 82: 34-55 (1974).

Ross, B. 1997. Eco-friendly procurement training course for UN HCR. : 126 p.

Ryan, S., and Skipworth, M., (1993), Consumers turn their backs on green revolution, The Times, 4 April

Salzman, J. 1997. Informing the Green Consumer: The Debate over the Use and Abuse of Environmental Labels. Journal of Industrial Ecology 1 (2): 11-22.

Sanders, W. 1997. Environmentally Preferable Purchasing: The US Experience (946-960) in Greener Purchasing: Opportunities and Innovations. Sheffield, Greenleaf Publ. 325p.

Sayre, D. 1996. Inside ISO 14000: The competitive advantage of environmental management. Delray Beach FL., St. Lucie Press. 232p.

SHAPIRO, C. Premiums for High Quality Products as Returns to Reputa- tion. Quarterly Journal of Economics, Vol. 98, No. 4 (1983), pp. 659-680.

Stillwell, M. and van Dyke, B. 1999. An activists handbook on genetically modified organisms and the WTO. Washington DC., The Consumer's Choice Council: 20 p.

Semenzato, A.; Costantini, A.; Meloni, M.; Maramaldi, G.; Meneghin, M.; Baratto, G. Formulating O/W Emulsions with Plant-Based Actives: A Stability Challenge for an Eective Product. Cosmetics 2018, 5, 59.

Teisl, M. F., B. Roe, and R. L. Hicks. "Can Eco-labels tune a market? Evidence from dolphin-safe labeling," Presented paper at the 1997 American Agricultural Economics Association Meetings, Toronto.

THE GERSEN, C. Psychological Determinants of Paying Attention to Eco- Labels in Purchase Decisions: Model Development and Multinational Vali- dation. Journal of Consumer Policy, Vol. 23, No. 4 (2000), pp. 285-313.

Tibor, T. and Feldman, I. 1995. ISO 14000: a guide to the new environmental management standards. Burr Ridge Ill., Irwin Professional Publ. 250 p.

Torre, I. de la, & Batker, D. K. (n.d.) 1999-2000. Prawn to trade: prawn to consume. Graham WA., Industrial Shrimp Action Network (isatorre@seanet.com), [and] Asia –Pacific

Townsend, M. 1998. Making things greener: motivations and influences in the greening of manufacturing. Aldershot, England, Ashgate Publisher. 203p.

U.S. Energy Information Administration, What is U.S. Electricity Generation by Energy Source?, Retrieved From: https://www.eia.gov/tools/faqs/faq.php?id=427&t=3

U.S. Energy Information Administration, Biomass Explained, Retrieved From: https://www.eia.gov/energyexplained/?page=biomass_home

U.S. Environmental Protection Agency. National Water Quality Fact Inventory: 1990 Report to Congress. EPA 503-9-92-006, Apr. 1992.

UK Eco-labelling Board website, accessed via http://www.ecosite.co.uk/Ecolabel-UK/

US Environmental Protection Agency (EPA742-R-99-001): 40 p. <www.epa.gov/opptintr/epp>

US EPA, 1993. Determinants of effectiveness for environmental certification and labeling programs. Washington, D.C., US Environmental Protect

US EPA, 1993. Status report on the use of environmental labels worldwide. Washington, D.C., US Environmental Protection Agency (742-R-93-001 September).

US EPA, 1993. The use of life-cycle assessment in environmental labeling. Washington, D.C., US Environmental Protection Agency (742-R-93-003 September).

US EPA, 1998. Environmental labeling: issues, policies, and practices worldwide. Washington DC., Environmental Protection Agency, Pollution Prevention Division Prepared by Abt

US EPA, 1999. Comprehensive procurement guidelines (CPG) program. Washington, D.C., US Environmental Protection Agency: <www.epa.gov/cpg>

US EPA, 1999. Environmentally preferable purchasing program: Private sector pioneers: How companies are incorporating environmentally preferable purchases. Washington, D.C.,

USG, 1993. Federal acquisition, recycling, and waste prevention. Washington DC., Executive Order: (20 October).

USG, 1998. Greening the government through waste prevention, recycling, and federal acquisition. Washington, D.C., Executive Order 13101 (September).

Kijjoa, A.; Sawangwong, P. Drugs and Cosmetics from the Sea. Mar. Drugs 2004, 2, 73–82. [CrossRef]

Wang, J.; Pan, L.; Wu, S.; Lu, L.; Xu, Y.; Zhu, Y.; Guo, M.; Zhuang, S. Recent Advances on Endocrine Disrupting Eects of UV Filters. Int. J. Environ. Res. Public Health 2016, 13, 782.

Bilal, A.I.; Tilahun, Z.; Shimels, T.; Gelan, Y.B.; Osman, E.D. Cosmetics Utilization Practice in Jigjiga Town, Eastern Ethiopia: A Community Based Cross-Sectional Study. Cosmetics 2016, 3, 40.

Ting, C.T.; Hsieh, C.M.; Chang, H.-P.; Chen, H.-S. Environmental Consciousness and Green Customer Behavior: The Moderating Roles of Incentive Mechanisms. Sustainability 2019, 11, 819.

Chen, K.; Deng, T. Research on the Green Purchase Intentions from the Perspective of Product Knowledge. Sustainability 2016, 8, 943.

Wang, H.; Ma, B.; Bai, R. How Does Green Product Knowledge Eectively Promote Green Purchase Intention? Sustainability 2019, 11, 1193.

Nguyen, T.T.H.; Yang, Z.; Nguyen, N.; Johnson, L.W.; Cao, T.K. Greenwash and Green Purchase Intention: The Mediating Role of Green Skepticism. Sustainability 2019, 11, 2653.

Cinelli, P.; Coltelli, M.B.; Signori, F.; Morganti, P.; Lazzeri, A. Cosmetic Packaging to Save the Environment: Future Perspectives. Cosmetics 2019, 6, 26.

Eixarch, H.; Wyness, L.; Siband, M. The Regulation of Personalized Cosmetics in the EU. Cosmetics 2019, 6, 29.

Appendix I: Search by Logos

Here you can search the logos in this volume. It will help you to better undersand the Ecolabels you may encounter while shopping. Buying Eco-products will aid in having a better environment with minimum polution during production processes. Three important parameteres for shopping are **quality**, **price** & **environmental impacts** of the products.

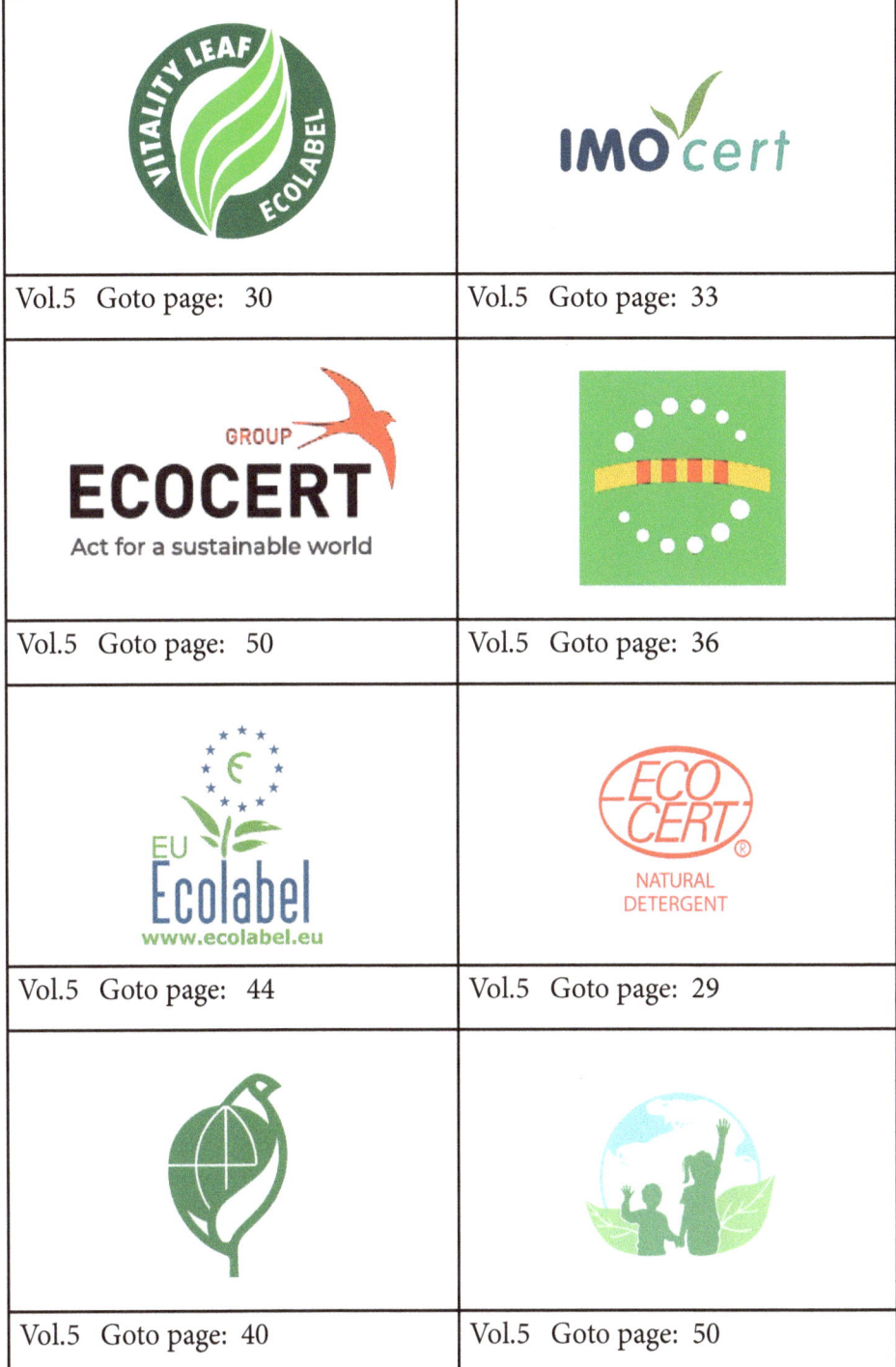

INTERNATIONAL ENVIRONMENTAL LABELLING VOL.5

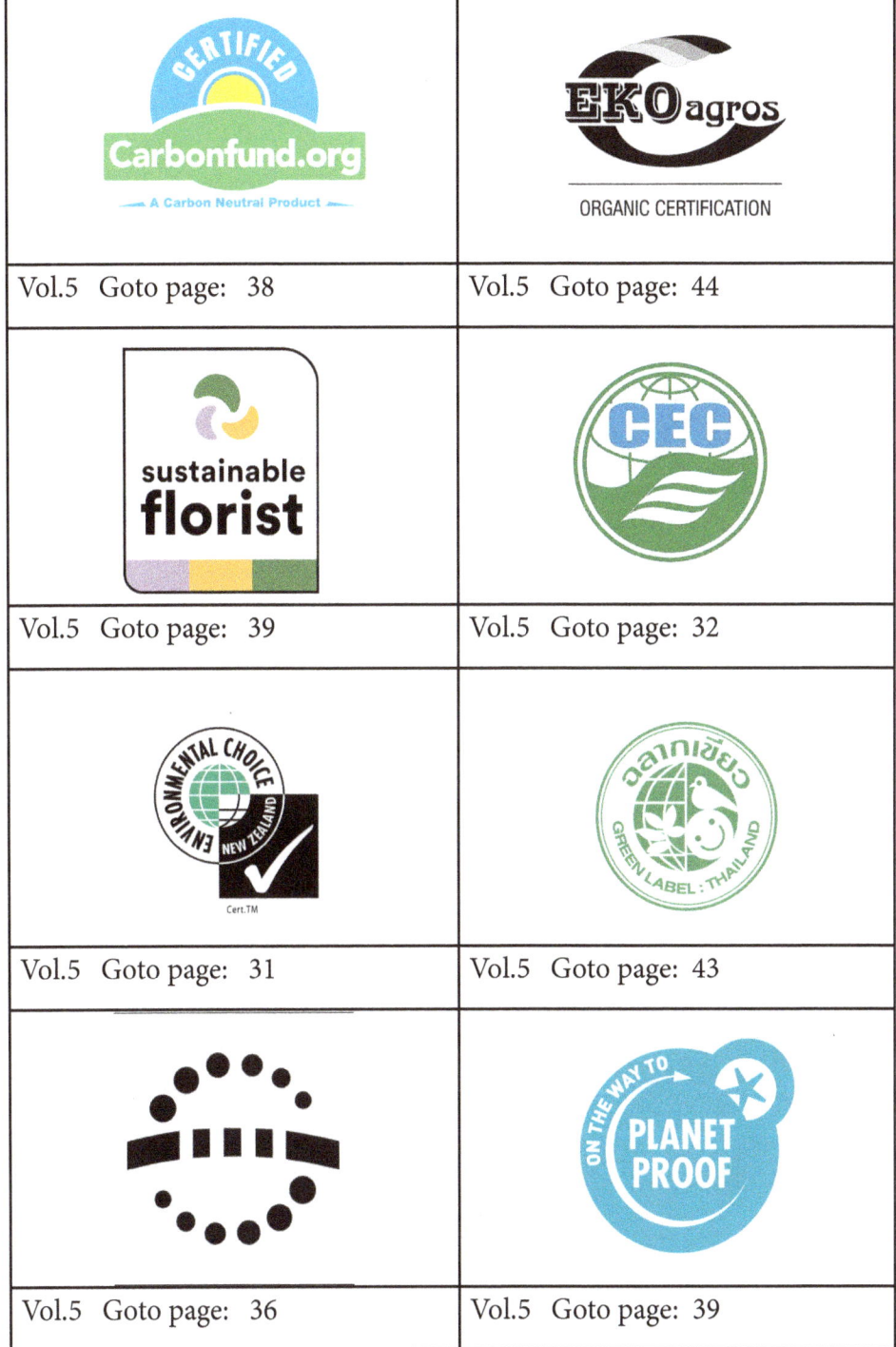

Appendix II

Using Algae to Clean Wastewater

Algae—as a renewable feedstock—grow a lot quicker than crops of corn or soybeans, We can start a new batch of algae about every seven days. It's a more continuous source that could offset 50 percent of our total gas use for equipment that uses diesel. Algae will take out all the ammonia and more than 80 percent of the nitrate and near 100 percent of the phosphate from the wastewater.

Algae can be used in wastewater treatment for a range of purposes, some of which are used for the removal of coliform bacteria, reduction of both chemical and biochemical oxygen demand, removal of N and/or P, and also for the removal of heavy metals.

Biodegradation is the most effective way by which microalgae eliminate organic contaminants from an aqueous phase. ... In one example of biodegradation of progesterone by two freshwater microalgal species, S. obliquus and C. pyrenoidosa showed 95% biodegradation of the available progesterone in an aqueous medium. Microalgae form the viable biodiesel feedstock. They can be found in soils, ice, lakes, rivers, hot springs, and oceans, anywhere sunlight and water cooccur. They have a simple cellular structure and are a diverse group of photosynthetic eukaryotes ranging from unicellular to multicellular forms.

APPENDIX III

Environmental Friendly Photos

Environmental friendly photos will be placed in this appendix. These photos can be received in the Top Ten Award International Network inbox from anywhere and everywhere, all over the globe. You can send your appropriate photos to us for them to be considered for publishing in one of the future, related volumes. They will be published with proper credit to the sender. The pictures can also be images of the Ecolabels existing in products within your country.

INTERNATIONAL ENVIRONMENTAL LABELLING VOL.5 • 103

	# Vol.1 For All People who wish to take care of Climate Change, Food Industries: (Meat, Beverage, Dairy, Bakeries, Tortilla, Grain and Oilseed, Fruit and Vegetable, Seafood, And Sugar and Confectionery)
	# Vol.2 For All People who wish to take care of Climate Change, Electrical Industries: (Renewable Energy, Biofuels, Solar Heating & Cooling, Hydroelectric Power, Solar Power, Wind Power, Energy Conservation, Geothermal and Nuclear Power)
	# Vol.3 For All People who wish to take care of Climate Change, Fashion & Textile Industries: (Fashion Design, The Fashion System, Fashion Retailing, Marketing and Marchandizing, Textile Design and Production, Clothing and Textile Recycling)
	# Vol.4 For All People who wish to take care of Climate Change, Health & Beauty Industries: (Fragrances, Makeup, Cosmetics, Personal Care, Sunscreen, Toothpaste, Bathing, Nailcare & Shaving, Skin Care, Foot Care, Hair Care and Other Health & Beauty Products)

	## Vol.5 For All People who wish to take care of Climate Change, Maintenance & Cleaning Products: (All-purpose Cleaners, Abrasive Cleaners, Powders. Liquids, Specialty Cleaners, Kitchen, Bathroom, Glass and Metal Cleaners, Bleaches, Disinfectants and Disinfectant Cleaners)
	## Vol.6 For All People who wish to take care of Climate Change, Wood & Stationery Industries: (Wooden Products, Cardboard, Papers, Markers, Pens, NoteBooks, Writing Pads and Writing Sets, Pencils, White Papers, Envelopes and Organizers, Staplers and Paper Clips)
	## Vol.7 For All People who wish to take care of Climate Change, DIY & Construction Industries: (Do it yourself " ("DIY") of Building, Modifying, or Repairing, Renovation, Construction Materials, Cement, Coarse Aggregates. Clay Bricks, Power Cables, Pipes and Fittings, Plywood, Tiles, Natural Flooring)
	## Vol.8 For All People who wish to take care of Climate Change, Agricurture & Gardening Industries: (Shifting Cultivation, Nomadic Herding, Livestock Ranching, Commercial Plantations, Mixed Farming, Horticulture, Butterfly Gardens, Container Gardening, Demonstration Gardens, Organic Gardening)

Vol.9
For All People who wish to take care of Climate Change, Professional Products & Services: (Teachers, Pilots, Lawyers, Advertising Professionals, Architects, Accountants, Engineers, Consultants, Human Resources Specialist, R&D, Psychologists, Pharmacist, Commercial Banker, Research Analyst)

Vol.10
For All People who wish to take care of Climate Change, Financial Products & Services: (Banking, Professional Advisory, Wealth Management, Mutual Funds, Insurance, Stock Market, Treasury/Debt Instruments, Tax/Audit Consulting, Capital Restructuring, Portfolio Management)

Vol.11
For All People who wish to take care of Climate Change, Tourism Industries: (Airline Industry, Travel Agent, Car Rental, Water Transport, Coach Services, Railway, Spacecraft, Hotels, Shared Accommodation, Camping, Bed & Breakfast, Cruises, Tour Operators)

Set Box Books Vol.1-11 + Free Knowledge Test
for
Schools, Libraries, Homes and Offices all over the globe:

www.TopTenAward.Net

www.ingramcontent.com/pod-product-compliance
Lightning Source LLC
Chambersburg PA
CBHW040107120526
44589CB00039B/2777